Bibliographic information published by the German National Library:

The German National Library lists this publication in the National Bibliography; detailed bibliographic data are available on the Internet at http://dnb.dnb.de .

Imprint:

Print and binding: Books on Demand GmbH, Norderstedt Germany
ISBN: 978-3-668-05719-7

This book at GRIN:

http://www.grin.com/en/e-book/306806/a-critical-review-of-tayfur-s-sediment-transport-capacity-model

Muhammed Ernur Akiner

A critical review of Tayfur's sediment transport capacity model

GRIN Publishing

Critical review on sediment transport capacity model

Muhammed Ernur AKINER[*, 1]

[1]Akdeniz University, Vocational School of Technical Sciences, Campus, Antalya, Turkey

Summary

A paper is critically reviewed in this study. This paper is related with the sediment transport capacity models:

Critical review of "Applicability of Sediment Transport Capacity Models for Nonsteady State Erosion from Steep Slopes": This paper has been written by Tayfur (2002).

The physics-based sediment transport equations are derived from the assumption that the sediment transport rate can be determined by a dominant variable such as flow discharge, flow velocity, slope, shear stress, stream power, and unit stream power. In modeling of sheet erosion/sediment transport, many models that determine the transport capacity by one of these dominant variables have been developed. The developed models mostly simulate steady-state sheet erosion. Few models that are based on the shear-stress approach attempt to simulate nonsteady state sheet erosion. However, to the knowledge of the writer, there is no study that qualitatively investigated the applicability of all of the commonly employed sediment transport models in estimating sediment loads from steep slopes under nonsteady state conditions.

The objective of this study is to investigate the applicability of the transport capacity models that are based on one of the commonly employed dominant variables—unit stream power, stream power, and shear stress—to simulate nonsteady state sediment loads from steep slopes under different rainfall intensities.

Keywords: Erosion, modelling, nonsteady state, sediment transport, steep slopes.

The sediment transport capacity:

Most physics-based sediment transport equations were derived from the assumption that the sediment transport capacity could be determined by a dominant variable such as flow discharge, flow velocity, slope, shear stress, stream power, and unit stream power. The sediment transport capacity is expressed by the basic form

$$T_c = \eta_i (D - D_c)^{ki} \tag{1}$$

Where T_c =transport capacity (M/L/T); η_i and k_i are parameters related to flow and sediment conditions; D = dominant variable; D_c = critical condition of dominant variable at incipient motion.

Flow dynamics:

Kinetic wave approximation (KWA) is used for modeling nonsteady state flow dynamics in one dimension (1-D). Since this study focuses on sediment transport from steep slopes, KWA in one dimension is a fairly good approximation to the full Saint-Venant equations. The KWA equation in one dimension is stated as

$$\frac{\partial h}{\partial t} + \frac{\partial}{\partial x}(K h^{5/3}) = (r - i) \tag{2}$$

where

$$K = \frac{\sqrt{S}}{n} \tag{3}$$

and where h = overland flow depth (L); r = rainfall intensity (L/T); i = infiltration rate (L/T); S = bed slope; and n = Manning`s roughness coefficient ($L^{1/3}$/T).

Erosion dynamics:

The physics-based one-dimensional nonsteady state erosion/sediment transport equation can be expressed as (Li 1979; Woolhiser et al. 1990)

$$\frac{\partial (hc)}{\partial t} + \frac{\partial}{\partial x}(qc) = \frac{1}{\rho_s}(D_{rd} + D_{fd}) \tag{4}$$

where

$$q = K h^{5/3} \tag{5}$$

and where c = sediment concentration by volume (L^3/L^3); ρ_s = sediment particle density (M/L^3); q = unit flow discharge (L^2/T); D_{rd} = soil detachment rate by raindrops (M/L^2/T); and D_{fd} = soil detachment/deposition rate by sheet flow (M/L^2/T).

Soil detachment by raindrops:

Soil detachment is a function of the erosivity of rainfall and the erodibility of the soil particles. On a bare soil surface, detachment by raindrops can be expressed as (Li 1979)

$$D_{rd} = \alpha r^{\beta}\left(1 - \frac{z_w}{z_m}\right) \qquad (6)$$

where α = soil detachability coefficient, which depends on the soil characteristics ($M/L^2/L$). Sharma et al. (1993) obtained the range of 0.0006–0.0086 $kg/m^2/mm$ for α for easily detachable soils and 0.00012–0.0017 $kg/m^2/mm$ for less detachable soils. Note that in Eq. (6) r is in millimeters per hour, α is in kilograms per meter squared per millimeter, and D_{rd} is in kilograms per meter squared per hour. The parameter β is an exponent whose range is 1.0–2.0. From experimental studies, it is shown that $\beta = 2.0$ (Meyer 1971; Foster 1982). Sharma et al. (1993) showed that the value of β is in the range of 1.09–1.44. Foster et al. (1977) used a value of β of 1.0. Tayfur (2001) showed that the change in the value of β in between 1.0 and 1.8 does not affect the sediment discharge significantly. In the present study, the value of b is taken as 1.0.

Parameter z_w is the flow depth plus the loose soil depth (L), and z_m is the maximum penetration depth of raindrop splash (L). Eq. (6) is valid when $z_w < z_m$; otherwise, there is no detachment by the raindrops. According to Li (1979)

$$z_m = 3(2.23 r^{0.182}) \qquad (7)$$

Note that in Eq. (7), r is in millimeters per hour and z_m is computed in millimeters.

Eq. (6) expresses the detachment by raindrop impact as a power function of rainfall intensity, flow depth, and loose soil depth. As the sum of the flow depth and loose soil depth increases, the penetration depth decreases and consequently the detachment by raindrops decreases.

Soil detachment/deposition by sheet flow:

The soil detachment/deposition rate is proportional to the difference between the sediment transport capacity and the sediment load in the flow. When the sediment load is greater than the transporting capacity, deposition occurs. The soil detachment/deposition by sheet flow can be expressed as (Foster 1982; Govindaraju and Kavvas 1991).

$$D_{fd} = \varphi(T_c - q_s) \tag{8}$$

where

$$q_s = \rho_s c q \tag{9}$$

where q_s = unit sediment discharge (M/L/T). If the transport capacity exceeds the existing unit sediment discharge ($T_c > q_s$), the flow will detach particles; otherwise, it will deposit the particles. Parameter φ is the transfer rate coefficient (1/L), which may vary over a wide range, depending upon the soil type. Foster (1982) gives the range φ = 3–33 m^{-1} for sand. In the present study, during detachment ($T_c > q_s$), φ is taken as 24 m^{-1}. During deposition ($T_c < q_s$), φ is estimated as a function of particle terminal fall velocity (V_f) and the unit flow discharge (q) as (Foster 1982)

$$\varphi = (0.5V_f)/q \tag{10}$$

Yang (1996) expresses the particle terminal fall velocity (V_f) as a function of the particle diameter and the particle Reynolds number. The particle Reynolds number can be expressed as (Woolhiser et al. 1990)

$$\mathscr{R}_{pn} = \frac{V_f d}{\upsilon} \tag{11}$$

where (V_f) = particle terminal fall velocity (L/T); $\mathscr{R}_{pn}$= particle Reynolds number; d = particle diameter (L)); and υ = kinematic viscosity of water (L^2/T). When the particle Reynolds number ($\mathscr{R}_{pn}$) is less than 2.0, the terminal fall velocity of a particle is expressed as (Yang 1996)

$$V_f = \begin{cases} \dfrac{1}{18}\dfrac{(\gamma_s - \gamma)}{\gamma}\dfrac{gd^2}{\upsilon}; & d \leqslant 0.1 \text{ mm} \\ F\left[\dfrac{gd(\gamma_s - \gamma)}{\gamma}\right]^{0.5}; & 0.1 \text{ mm} < d \leqslant 2.0 \text{ mm} \\ 3.32\sqrt{d}; & d > 2.0 \text{ mm} \end{cases} \tag{12}$$

where g = gravitational acceleration (L/T^2); γ_s = specific weight of sediment ($M/L^2/T^2$); and γ = specific weight of water ($M/L^2/T2$); and

$$F = \begin{cases} \left[\dfrac{2}{3} + \dfrac{36\upsilon^2\gamma}{gd^3(\gamma_s - \gamma)}\right]^{0.5} - \left[\dfrac{36\upsilon^2\gamma}{gd^3(\gamma_s - \gamma)}\right]^{0.5}; & 0.1 \text{ mm} < d \leqslant 1.0 \text{ mm} \\ 0.79; & 1.0 \text{ mm} < d \leqslant 2.0 \text{ mm} \end{cases} \tag{13}$$

Note that in Eq. (12), V_f is in meters per second and d is in meters. When the particle Reynolds number is greater than 2.0, the terminal fall velocity is determined experimentally. Yang (1996) gives a figure summarizing the fall velocity values depending on the sieve diameter and the shape factor. For most natural sands, the shape factor is 0.7. Rouse (1938) gives $V_f = 0.024$ m/s for $d = 0.2$ mm. In the present study, for $\mathscr{R}_{pn} > 2.0$, the terminal fall velocity is assumed to be 0.024 m/s.

Transport capacity models:

Sheet flow transport capacity is a function of several factors that include runoff rate, flow velocity, slope steepness of the surface, transportability of detached soil particles, and the effect of raindrop impact. The basic relationship that does not take into account the effect of raindrop impact on the transport capacity might be a typical sediment transport equation form of Eq. (1). Depending upon the chosen model for the sediment transport capacity of sheet flow, the dominant variable can be shear stress, stream power, and the unit stream power. In the following sections, a brief description of each approach is given.

Shear stress approach:

The transport capacity model that is based on the dominant variable shear stress can be expressed as (Foster 1982; Govindaraju and Kavvas 1991; Yang 1996)

$$T_c = \eta_\tau (\tau - \tau_c)^{k_\tau} \tag{14}$$

where

$$\tau = \gamma h S \tag{15}$$

and where τ = shear stress, which is the tractive force developed by the sheet flow to overcome the critical shear stress ($M/L/T^2$); and η_i = soil erodibility coefficient, which is a function of particle diameter and density. While its value may vary over a wide range, Foster (1982) suggests the value of 0.6 for η_i . Parameter k_i = exponent whose value varies between 1 and 2.5. Foster (1982) suggests the value of 1.5 for k_i . Parameter τ_c = critical shear stress ($M/L/T^2$), which is a function of the particle diameter and specific weight of the sediment and water. Li (1979) expresses τ_c as

$$\tau_c = \delta_s (\gamma_s - \gamma) d \tag{16}$$

where δ_s = a constant dependent on flow conditions. Gessler (1965) shows that ds should be 0.047 for most flow conditions. If rilling develops on the overland flow surface, the value of δ_s

should be lower (Li 1979). Parameter τ_c represents the resistance of the soil against erosion. The critical shear stress is very small for cohesionless soils, and it is often neglected (Foster 1982).

Stream power approach:

The transport capacity model that is based on the dominant variable stream power can be expressed as (Yang 1996)

$$T_c = \eta_{\tau v}(\tau V - \tau_c V_c)^{k_{\tau v}} \tag{17}$$

where V = flow velocity (L/T); and V_c = critical flow velocity at incipient sediment motion (L/T). In Eq. (17), τV = stream power and $\tau_c V_c$ = critical stream power at incipient sediment motion. V is computed from the flow dynamics part of the model as

$$V = Kh^{2/3} \tag{18}$$

Yang (1996) expresses the critical flow velocity as being dependent upon the shear velocity Reynolds number. The shear velocity Reynolds number is expressed as

$$\mathscr{R}^* = \frac{u_* d}{\upsilon} \tag{19}$$

where u_* = shear velocity (L/T) and is defined as (Yang 1996)

$$u_* = \sqrt{ghS} \tag{20}$$

The critical flow velocity at incipient sediment motion is expressed as (Yang 1996)

$$V_c = \begin{cases} \dfrac{2.5V_f}{\log(\mathscr{R}^*) - 0.06} + 0.66V_f; & 1.2 < \mathscr{R}^* < 70 \\ 2.05V_f; & \mathscr{R}^* > 70 \end{cases} \tag{21}$$

Unit Stream power concept:

The transport capacity model that is based on the dominant variable unit stream power can be expressed as (Yang 1996)

$$T_c = \eta_{vs}(VS - V_c S_c)^{k_{vs}} \tag{22}$$

where S = energy slope, which is assumed to be equal to the bed slope; and S_c = critical slope at incipient sediment motion. In Eq. (22), VS = unit stream power; and $V_c S_c$ = critical unit stream power at incipient sediment motion. By utilizing Meyer-Peter and Muller's (1948) bed load equation, the slope at incipient motion (S_c) can be obtained as

$$S_c = \frac{0.058dn^{1.5}}{hd_{90}^{0.25}} \tag{23}$$

where d_{90} = bed material size, where 90% is finer (L). Note that in Eq. (23), h, d, and d_{90} are in meters.

Solution procedure:

Eqs. (2) and (4) were solved numerically by using the implicit centered finite difference method. The Newton-Raphson iterative technique was used to solve the set of nonlinear equations resulting from the implicit procedure. As upstream boundary conditions, zero flow depth and zero sediment concentration were used. As downstream boundary conditions, zero depth gradient and zero sediment concentration gradient were employed. Since rainfall starts on a dry surface, there is initially no flow and erosion on the hillslope surface. Under the specified initial and boundary conditions, the numerical solutions of Eqs. (2) and (4) are executed simultaneously for each time step. Every time step, Eq. (2) is first solved to obtain flow depths, flow velocities, and unit flow discharges. Then Eq. (4) is solved to compute sediment concentrations and unit sediment discharges. Every time step (j), the loose soil depth (l_d), which is required by Eq. (6), is also computed (reminding: z_w is the flow depth plus the loose soil depth (L)).

The loose soil depth:

The loose soil depth at the (j+1) time step is computed as

$$l_d(j+1) = l_d(j) - [D_{rd}(j) + D_{fd}(j)]\frac{\Delta t}{\rho_s} \tag{24}$$

Δt = time step (T); ρ_s = mass density of sediment particles (M/L^3).

Analysis of results:

Kilinc and Richardson (1973) performed experimental studies by using a 1.21 m high x 1.52 m wide x 4.58 m long flume with an adjustable slope. Commercial sprinklers on 3 m risers, placed 3 m apart along the sides of the flume, simulated rainfall. The flume was filled with compacted sandy soil (90% sand and 10% silt and clay), which was leveled and smoothed before each run. The soil had a nonuniform size distribution with d_{50} = 0.35mm (the median diameter of the sediment recorded for 50% of the samples having a diameter finer than this size! and d_{90} = 1.3 mm. The compacted sandy soil had a bulk density of 1,500 kg/m^3 and a porosity of 0.43. The major controlled variables were rainfall intensity and soil surface slope. Infiltration and erodibility of the surface were constant. On average, the constant infiltration rate for each run was about 5.3 mm/h. The calibrated values of the model parameters that resulted in the best fit for the observed experimental data are as follows:

• Manning's roughness coefficient (η): 0.012 ($m^{1/3}/s$),

• Soil detachability coefficient (α): 0.0012 (kg/m2/mm),

• Soil erodibility coefficient ($\eta_{\tau}= \eta_{\tau v}= \eta_{vs}$): 0.10,

• (Unit stream power) exponent (k_{vs}): 1.56,

• (Stream power) exponent ($k_{\tau v}$): 1.18, and

• (Shear stress) exponent (k_{τ}): 1.92.

Values of the exponents (k_i) or (k_{vs} ,$k_{\tau v}$,k_{τ}) are different for each model. The models were calibrated by one of the data sets, and employed to simulate different sediment loads from different slopes (5.7, 10, 15, 30, and 40%) under two different rainfall intensities (57 and 93 mm/h).

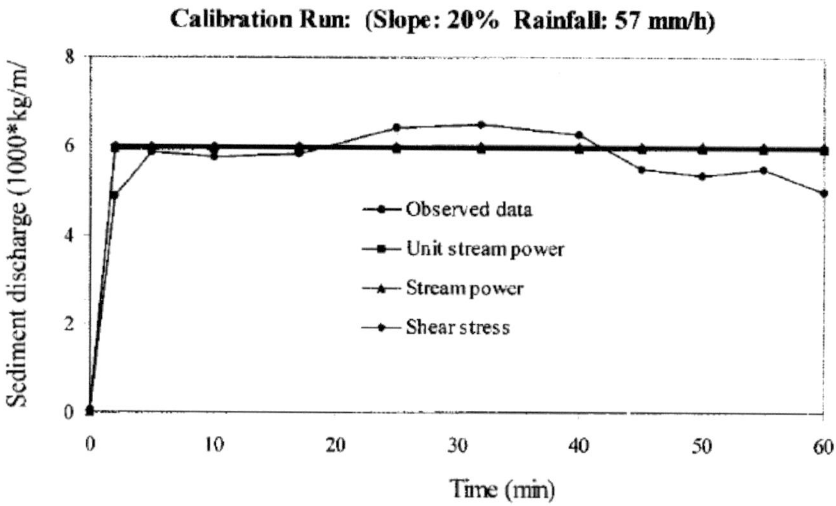

Figure a: Simulation of observed data; calibration run (S = 20%, r = 57 mm/h). Source: Tayfur (2002)

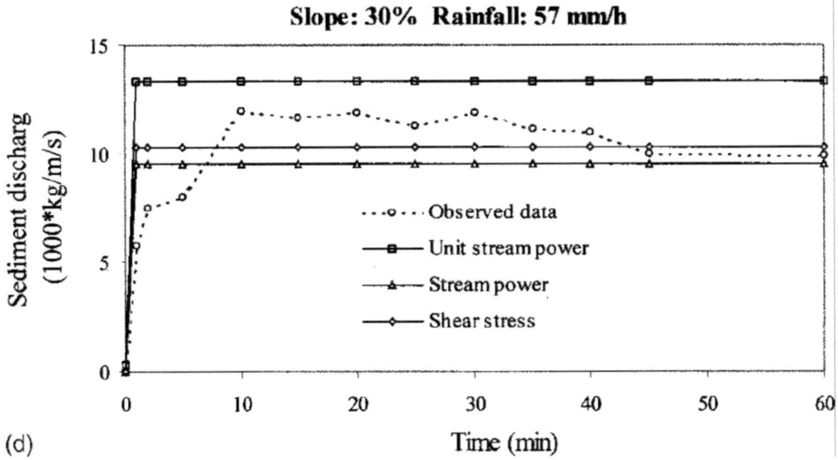

Figure b: Simulation of observed data; (S = 30%, r = 57 mm/h). Source: Tayfur (2002)

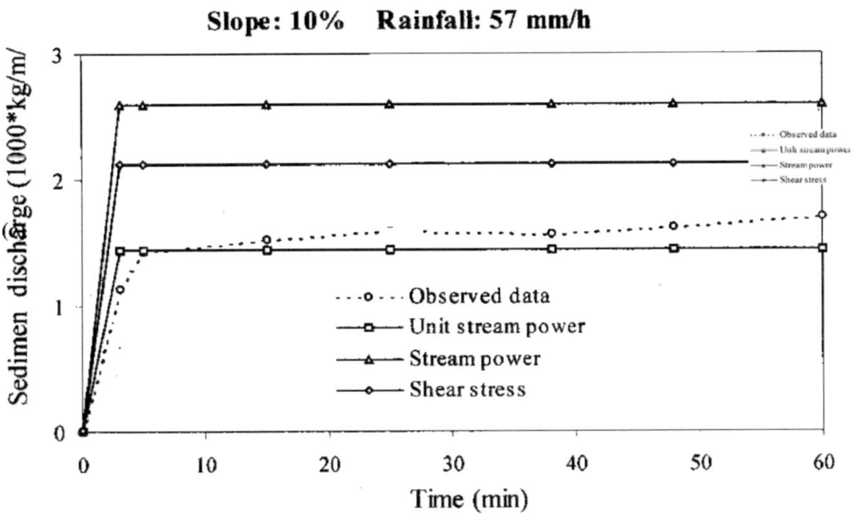

Figure c: Simulation of observed data; (S = 10%, r = 57 mm/h). Source: Tayfur (2002)

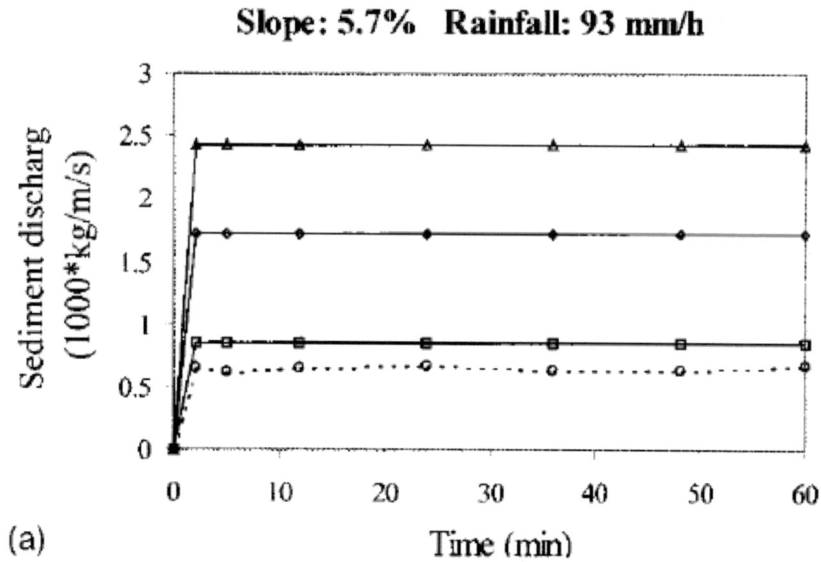

Figure d: Simulation of observed data; (S = 5.7%, r = 93 mm/h). Source: Tayfur (2002)

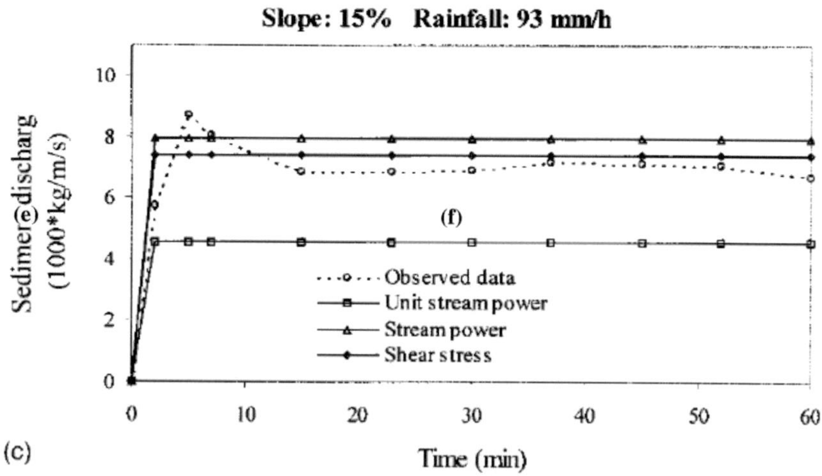

Figure e: Simulation of observed data; (S = 15%, r = 93 mm/h). Source: Tayfur (2002)

Slope: 40%   Rainfall: 93 mm/h

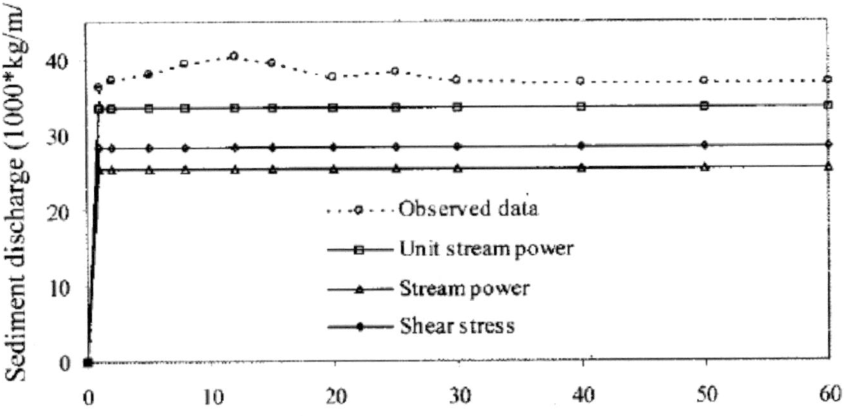

(e)

Figure f: Simulation of observed data; (S – 40%, r = 93 mm/h). Source: Tayfur (2002)

Conclusion and recommendations

The author presents a very interesting study by investigating the applicability of the previously mentioned erosion/sediment transport models for nonsteady state erosion from steep slopes. Nevertheless, there might be some additions to this study in order to improve the intelligibility and quality of the study. For instance we have knowledge about parameters used in the experimental study. However there is no figure for the depiction of the contrivance, experimental setup or equipments used. Author gives references for the experimental setup but it would be better if the author could give some explanatory schemes or pictures regarding experimental setup. This study is for infiltrating surfaces. Surfaces are rarely smooth by nature, and the constant slope assumption is far from realistic. Additionally in this study an average constant infiltration (5.3 mm/h) was used on the other hand time varying infiltration is more realistic than constant infiltration, since rainfall generally starts on dry, unsaturated soil (as same as we take into consideration during infiltration-rain intensity analysis with respect to Green and Ampt). The effect of varying roughness on total sediment discharge is relatively negligible, however as far as local erosion measures are concerned; varying roughness might be taken into account. A new study on related topic can be prepared by taking the above mentioned suggestions into account.

The models were found to be very sensitive to the changes in rainfall intensities and slopes. An increase/decrease in rainfall intensity and slope results in an increase/decrease in the sediment yield, and each model is able to capture this behavior. However, the performance of each model in simulating sediment yields from different slopes was found to be very much dependent upon the steepness of the slope, and the intensity of the rainfall. These are comparative performances of models with respect to each other. Therefore, the slope steepness and rainfall intensity play a major role in the selection of an appropriate sediment transport capacity model in simulating nonsteady state sediment loads by sheet flow.

The test of the calibrated models with observed data sets shows that the unit stream power model gives better simulation of sediment loads from mild slopes. The stream power and the shear stress models, on the other hand, simulate sediment loads from steep slopes more satisfactorily. The exponent (k_i) in the sediment transport capacity formula is found to be 1.2, 1.9, and 1.6 for the stream power model, the shear stress model, and the unit stream power model, respectively.

Consequently if the general performance and good aspects of this study is examined, it is seen that this paper provides a useful contribution to literature.

References

Foster, G. R. (1982). ''Modelling the erosion process.'' C. T. Haan et al., eds., *American Society of Agricultural Engineers*, St. Joseph, Mich., 295–380.

Gessler, J. (1965). ''The beginning of bedload movement of mixtures investigated as natural armouring in channels.'' *Rep.,* Laboratory of Hydraulics and Water Resources, CIT.

Govindaraju, R. S., and Kavvas, M. L. (1991). ''Modelling the erosion process over steep slopes: Approximate analytical solutions.'' *J. Hydrol.*, 127, 279–305.

Kilinc, M., and Richardson, E. V. (1973). ''Mechanics of soil erosion from overland flow generated by simulated rainfall.'' *Hydrology Paper 63*, Colorado State Univ., Fort Collins, Colo.

Li, R. M., 1979, Water and Sediment Routing from Watersheds, *Modeling of Rivers*, H. W. Shen (Ed.), John Wiley & Sons, New York, Chap. 9, 9-1 to 9-88.

Meyer-Peter, E., and Muller, R. (1948). ''Formula for bed load transport.'' *2nd Meeting, Int. Association for Hydraulic Structures*.

Sharma, P. P., Gupta, S. C., and Foster, G. R. (1993). ''Predicting soil detachment by raindrops.'' *Soil Sci. Soc. Am. J.*, 57, 674–680.

Tayfur, G. (2002). ''Applicability of sediment transport capacity models for nonsteady state erosion from steep slopes.'' *J. Hydrologic Eng.*, 7(3), 252–259.

Woolhiser, D. A., Smith, R. E., and Goodrich, D. C. (1990). ''KINEROS—A kinematic runoff and erosion model: Documentation and user manual.'' *ASR-77*, U.S. Dept. of Agriculture, Agriculture Research Service.

Yang, C. T. (1996). *Sediment transport theory and practice*, McGraw- Hill, New York.